The best of **Getaway**

GALLERY

Getaway Books
(a division of Ramsay, Son & Parker Pty Ltd)
3 Howard Drive
Pinelands
7405
Cape Town

First published in 2004

Publisher:
Stirling Kotze

Managing editor:
David Bristow

Written and compiled by:
Robyn Daly and Justin Fox

Photographic research and proofreading:
Margy Beves-Gibson

Designer:
Ryan Manning

Sub-editor:
Marion Boddy-Evans

Reproduction by Unifoto (Pty) Ltd, Cape Town
Printed and bound by Tien Wah Press (Pte) Ltd, Singapore

ISBN: 0 620 31523 7

Photo credits:
Front cover: Mud tile exposure – Hendrik Ferreira, Johannesburg (August 1999).
Back cover: Body art of the Karo Karo people, South Omo, Ethiopia – Regina Brierley,
Switzerland (December 2000).
Half title page: Winter no 2 – Anoulize Strydom, Bloemfontein (November 2003).
Title page: Namibian trio – Sally Swart, Rhode Island, USA (January 2001).
Contents page: Blue crane – Friedrich von Hörsten, Somerset West (April 2002).

The best of **Getaway**
GALLERY

15 years photographing Africa

Contents

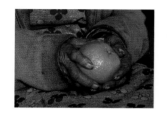

Welcome to *The Best of Getaway Gallery*, a collection of the finest photographs to appear in the magazine over the past 15 years. When Gallery was started with the first issue in 1989, the aim was to uncover new photographic talent. It did just that, and more. It proved to be a spawning ground for amateur photographers, many of whom had their images first published in Gallery, and some – the likes of Richard du Toit, Adrian Bailey, the Von Hörstens, Gerhard Dreyer and others – went on to become top practitioners in their field.

Over the years, *Getaway* magazine has become synonymous with the finest African photography and continues to pioneer new ways of seeing our continent. In this, the magazine owes as much to those who enter Getaway Gallery each month as it does to its portfolio contribu-tors and team of photojournalists. Every year the benchmark is raised as readers compete for monthly and annual photo-graphic prizes. The results are remarkable, as these pages attest. Accompanying the pictures is a brief description of how each photographer 'got the shot'. Some reveal the luck of being in the right place at the right time; others are the result of hours or days of painstaking preparation. Either way, every photograph here displays an empathy with the subject and an astute eye. Most of these photographers remain amateur enthusiasts who do not seek fame or fortune through their images. It is their love of nature and of photography that is reward in itself.

This book belongs to the readers of the magazine. It is their creation, and an inspira-tion to future Getaway Gallery generations.

People of Africa

Life, like the roads that crisscross our continent, imprints its map on the faces of Africa's people. Behind each of these pictures lies a narrative. Little details – the dirt under an old woman's fingernails – tell a greater story. The photographers have captured moments in the lives of Africans: ancient traditions, the marketplace, religious customs, rituals and people at play. The images invite us to share in the Ethiopian monk's collecting, the Xhosa boys' proud initiation, a 'ceremonial' braai or the thrill of catching a wave off Llandudno Beach. Mostly it's the everyday that has caught the photographer's eye – the human-ness of people. And it is the humanity of these pictures that makes them special.

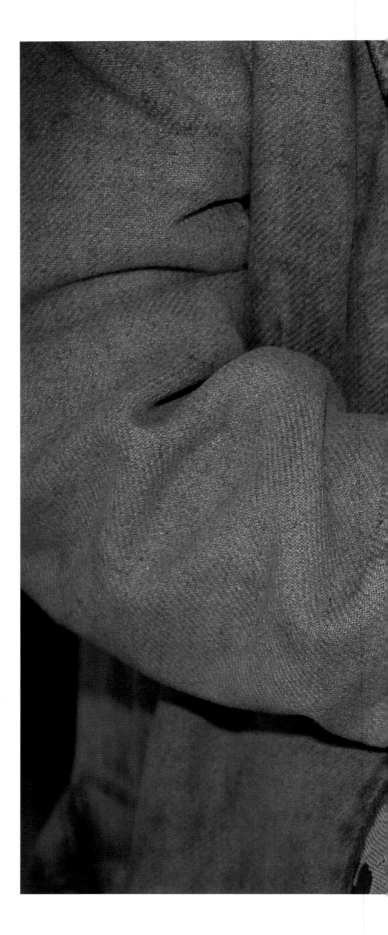

Ethiopian orthodox: Robert Brierley, Switzerland (April 2001).
The Orthodox Church is the centre of worship and community life in Ethiopia. This priest was conducting a collection on behalf of his parish at a roadside bus stop. The vibrant colour of his robe is what captured Robert's attention. The detailed set of rosary beads and bell added the signature to this image.

PREVIOUS PAGES
Embracing: Marike Bruwer, Piketberg (March 2000).
Visiting the village of Leliefontein, high in the Namaqualand mountains, Marike struck up a conversation with this old woman. She lived in a little reed hut with all her worldly belongings in one room. When Marike distributed oranges among the woman's grandchildren, a child handed his granny one. "I immediately saw our whole afternoon's conversation summed up in this picture." Marike underexposed by one stop and used fill-in flash.

Body art of the Karo Karo people, South Omo, Ethiopia: Regina Brierley, Switzerland (December 2000).

Of all the ethnic groups of Ethiopia's Lower Omo, the Karo excel in body art. They're also the most threatened tribe, numbering fewer than 1000. This warrior's willingness to be photographed certainly helped Regina. The picture makes a strong graphic impression, a powerful synecdoche for the clan's identity.

Outrigger canoes: Pete Oxford, Amanzimtoti (July 2000).

"I was a lecturer aboard the MV *World Discoverer*, an expedition cruise ship, when it called into Hellsville on Nosy Bé Island in northwest Madagascar. No sooner had we dropped anchor than a flotilla of outrigger canoes gathered round the ship where, from my high vantage point on the bridge wing, I was able to take the shot. Full of smiles, they had an eclectic range of wares for sale."

Llandudno boogie board: Geoff Spiby, Cape Town (July 1999).

Geoff took his son Kevin boogie boarding at Llandudno (Kevin emphasises that he is now a *real* surfer). With the camera in an underwater housing and a small strobe attached, Geoff waited in the water near to where Kevin was taking off, then swam alongside firing the shots as the wave took him. By using a wide-angle lens, Geoff could shoot from close and still get the mountains in.

Initiates: Friedrich von Hörsten, Somerset West (August 1999).

On an early-morning trip to Addo National Park Friedrich was travelling along an ugly, polluted section outside Port Elizabeth. The drive was pleasantly interrupted by the appearance of these two young Xhosa initiates, who were happy to pose with their newfound manhood.

The braai: Haggis Black, Adelaide (February 1997).
"'Getaway' is a wonderful word for all of us who need to escape our office-bound environment. And the week-end braai must rank as the most popular relaxation in Southern Africa." Haggis took this picture on a wind-less night, braai smoke lingering in the air and the spotlight and leaves creating a series of patterns. The four-second-exposure was made possible by lying on the grass using a bean bag and a wide-angle lens.

Optical illusion: Brian and Lindi Worsley, Cape Town (March 1996).
"We came across a Tonga man walking along a remote road in the Siabuwa District of northern Zimbabwe. We stopped to chat and gave him some cold water – he had been walking a long time but declined a lift. When he sucked on the cold water, his eyes watered and his cheeks hollowed as he drank – and we realised what an unusual luxury it must be to have water any cooler than room temperature."

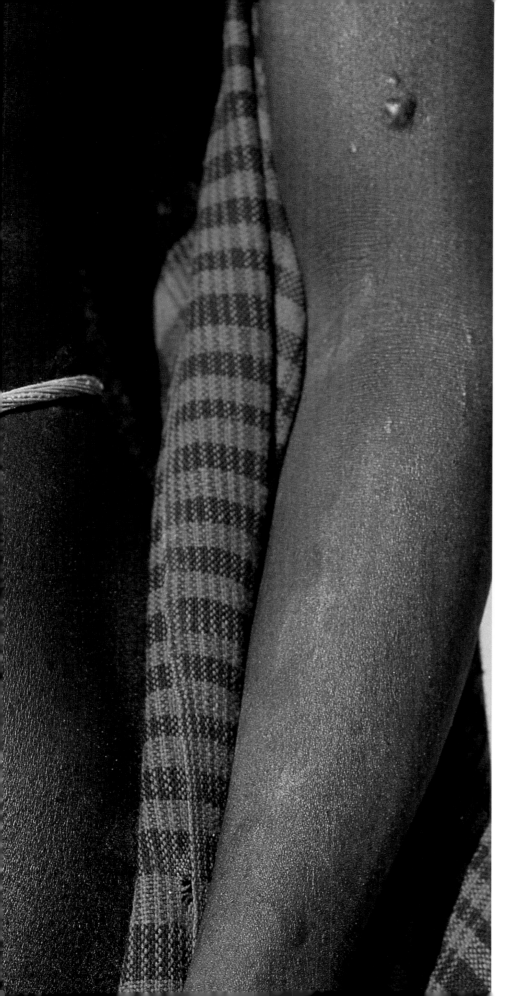

Surma, Kibbish: Robert Brierley, Switzerland (October 2001).

Kibbish is in southwestern Ethiopia, a truly wild region of the Lower Omo and home to the Surma people. Precious few visitors disembark from the Ethiopian Airlines Twin Otter here. Robert had come to walk for six days in the cauldron of the Lower Omo. "Warm evening light provided for a succession of photographs of this Surma tribesman with whom we completed the journey to Kibbish the following day."

**Early fishermen: Les Oates, Garsfontein
(March 1997).**

Les arrived at the Witbank Dam early one morning to get this shot of boys preparing to go fishing. The repetition – in the lines of the fishing rods, the silhouetted boys and their bicycles – is what makes the picture so successful.

**Klipvis: Tony Makin, Cape Town
(November 1993).**

There are 39 species of these ubiquitous little fish in the Western Cape. They occur commonly in rock pools and on shallow reefs and are very obliging subjects which often assume perfect poses. Tony used a Nikonos IVa with a 35 mm lens and a 2:1 extension tube at f16 and single Sea & Sea strobe. The film was Kodak Ekta ISO 25.

PREVIOUS PAGES
Jellyfish: Guido Zsilavecs, Cape Town (June 1999).

The *Maori* is a popular wreck dive site near Hout Bay. Guido's flash had developed a fault. Looking up, he noticed a school of jellyfish. The clear water and strong light made the subject irresistible and, being backlit, there was no need for fill-in flash. With much breath holding to avoid bubbles, the whole film was used in a short time.

Gas flames: Tony Makin, Cape Town (March 2000).

Gas flame nudibranchs are abundant in the shallow reefs of the Western Cape. Their cerata and vivid colours make stunning abstracts. This shot was taken with a housed Nikon F90, a 60 mm macro lens and a two times converter at f22 with twin Sea & Sea strobes.

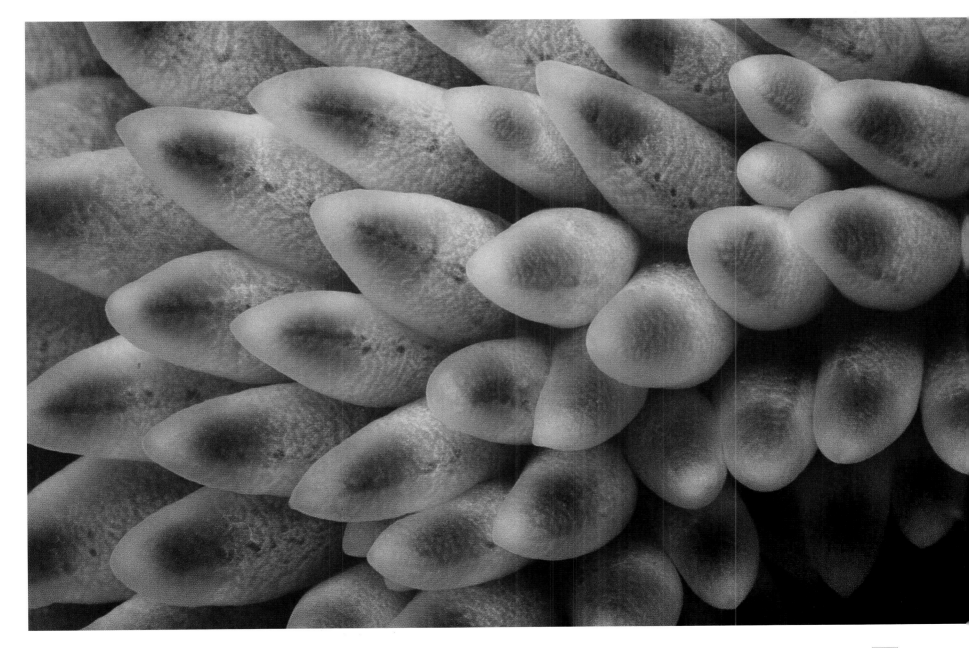

Fields of dreams: Peter Pinnock, Durban (October 1995).

In a typical Red Sea reef scene colourful soft corals abound. Schools of goldies explode out of the reef in a display of colour and beauty, retreating into crevices when approached. All sea goldies are born as females. The dominant male will spawn with a number of females in his 'harem'. If the male is taken by a predator, the most dominant female will change sex and take the place of the missing male.

PREVIOUS PAGES
H$_2$O and nothing else: Wally Shave, Cape Town (January 2000).

"Sailing back from Bazaruto, we were becalmed and forced to motor. It was a steaming hot day and the water was mirror flat, so we stopped to take a break. Given the conditions and the situation, it was perfect for a skinny-dip. The picture was taken on available light with no filters about 100 kilometres off Richard's Bay."

Let's go this way: Fiona Ayerst, Johannesburg (April 2004).

Banner fish school together for protection – grouping makes them seem bigger and there is also safety in numbers. These fish were hanging in mid-water about seven metres above a coral reef in Sodwana Bay. Schools of fish often scatter in a number of directions, making it difficult to capture them on film as a unit or to shoot anything other than their tails.

Red Sea snorkelling: Geoff Spiby, Cape Town (April 2004).

One windless day Geoff went snorkelling offshore near Sharm el Sheik. The sun's rays were penetrating deep and lighting up the soft corals. It was perfect for natural-light photography: all he did was position his companion for a pleasing composition.

Scorpion fish: Geoff Spiby, Cape Town (October 2002).

Diving off Ponta Malongane, Geoff's friend Greg Horn spotted this amazingly well-camouflaged scorpion fish. "Greg pointed at a large, yellow chunk of sponge which I thought was nice and gave it a glance. I then realised by his gesticulations that there was more to be seen and eventually spotted what all the excitement was about."

Goby on coral: Peter Pinnock, Durban (June 2001).
The goby can benefit from associating with a variety of hosts. Here a semi-transparent goby has made a head of hard coral its home. Other hosts may include coloured sponges, soft corals, sea cucumbers or gorgonian fans. These fish can vary in colour from semi-transparent to pale pink, yellow or green. They remain motionless on their host, making detection difficult.

Robber crab: Pete Oxford, Amanzimtoti (August 1998).

Politically part of the Seychelles, Aldabra Atoll lies remote from any landmass apart from a few sibling atolls. The Galapagos of the Indian Ocean, this single atoll is home to 10 times more giant tortoises than the entire Galapagos Archipelago. It is one of Pete's favourite places on earth. Robber crabs are known to cut coconuts from the tops of palm trees; on falling to the ground, the nuts split open and the crabs feed on them.

Partner shrimp and anemone: Dennis King, Umhlanga Rocks (November 2002).

The partner shrimp (*Periclimenes longgicarpus*) disguises itself by having a transparent body with white and violet markings, which matches the colours of its host, a bubble anemone (*Entacmaea quadricolor*). These tiny commensal shrimps – seen here in the Red Sea – live together in groups among the anemones' stinging tentacles. They've developed immunity to the stinging cells which grant them protection from predators. Partner shrimps are also occasionally known to clean fish by removing parasites.

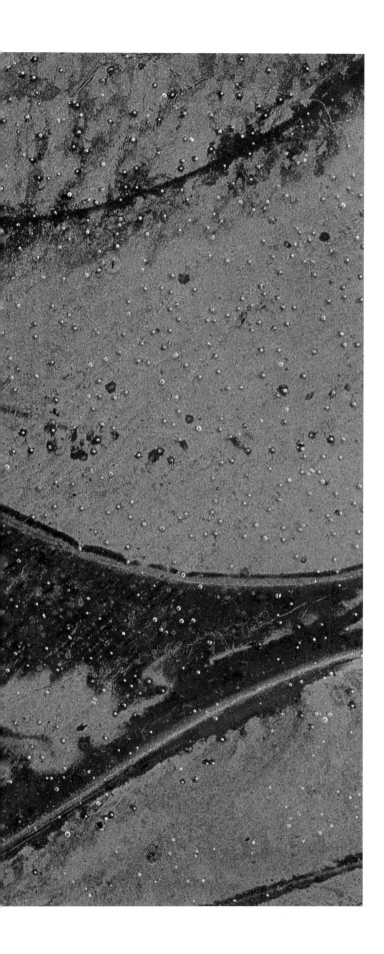

Patterns in nature

On a continent abundant with life, to narrow nature down to its bare elements is not easily achieved. But this is where abstract photographers thrive. They are, effectively, visual deconstructionists. It might involve traipsing through fields to find the two perfect sickle patterns 'in the haystack', peering closely at a sea anemone until a pleasing form emerges or manipulating the equipment to create new and captivating effects. Abstract photography is where *Getaway* gets closest to art photography, with images ranging from the precise details of water droplets to impressionistic multiple exposures. The images presented on these pages have startling clarity of vision and that special something that draws your eye back again and again.

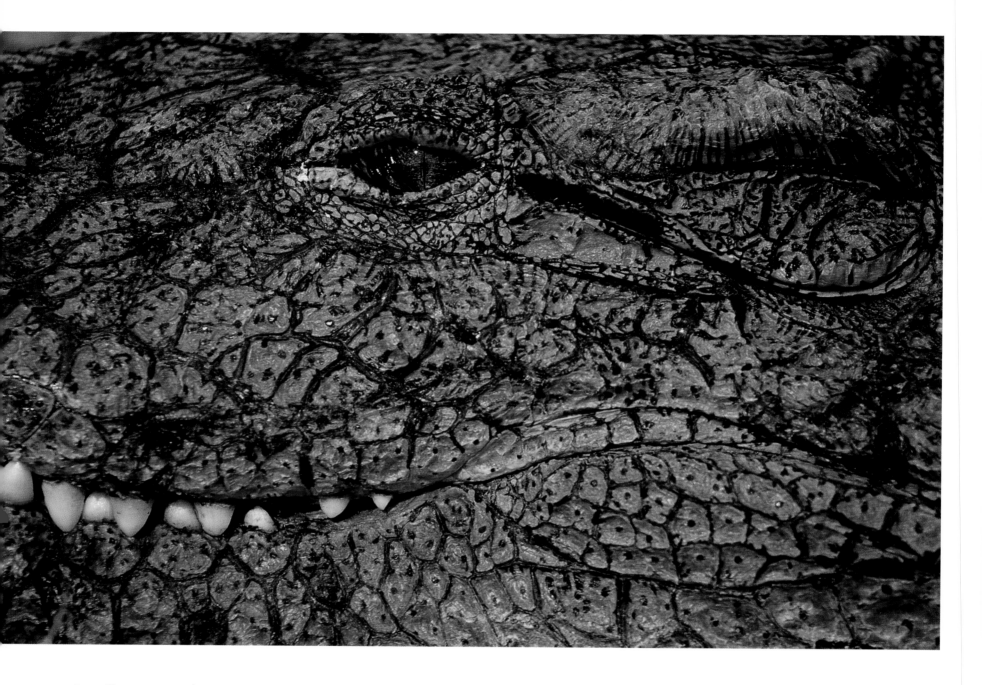

Crocodile pattern: Fred Hasner, Johannesburg (April 2002).

Fred had previously spotted this crocodile at Biyamiti Weir in the Kruger Park but found it impossible to get close. On this occasion he approached downhill, in reverse, with the engine off and camera with 500 mm lens on a beanbag. At the noise of the camera's shutter, the croc opened its eyes … and disappeared into the water.

PREVIOUS PAGES
Ice abstract: Geoff Spiby, Cape Town (December 2000).

A puddle of water on a gravel track at Driehoek in the Cedarberg had frozen overnight, trapping oak leaves and twigs in the ice. It's possible the puddle had frozen and semi-thawed a number times and cars had driven through it, creating these interesting patterns.

Earth forces: Hendrik Ferreira, Johannesburg (January 2002).

Natural light can transform the most mundane subject matter and is often considered the most important ingredient in photography. "Boulders are nature's three-dimensional sculptures and pose a special challenge to present effectively in two dimensions." This was taken in the Playground of the Giants, near Quiver Tree Forest, Keetmanshoop in Namibia. Hendrik used a Minolta X-700 camera and a Tamron 28–200 mm lens.

Circles of light: Peter Coqui, Port Elizabeth (April 2004).

A neighbour phoned Peter one day and told him to look up at the strange circles of light in the sky. "I did, and took 11 frames. But it was only when I read *Getaway*'s Indaba column (April 2003) that I found out what I had photographed – the halo being caused by light filtering through frozen cloud. One lamp post would have been boring, so I over-exposed a few shots and put them together."

Impressionistic tree: Marike Bruwer, Piketberg (March 1999).
Taken during a photographic workshop in Namaqualand, Marike shot nine multiple exposures on one frame while tilting the camera from a vertical to a horizontal position. She underexposed by three stops to compensate for the added light. "This image to me represents the feeling of spring."

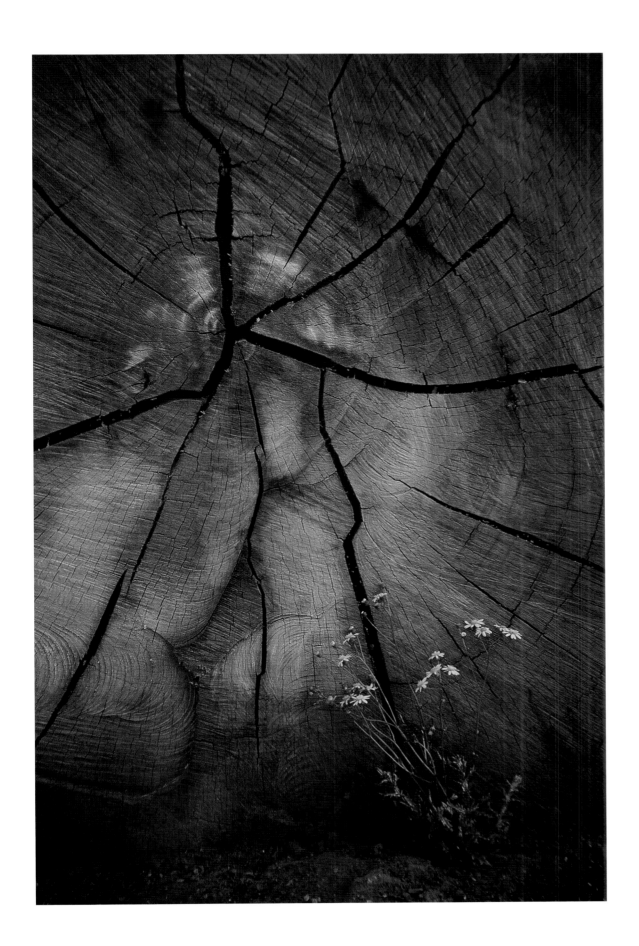

Slain forest giant: Francois le Roux, George (August 1996).

It is a sad irony that Francois' photograph was published in *Getaway* just a month before he died in 1996. His wife assured us when we told her it would be included in this collection that he had had the pleasure of seeing it in print.

OPPOSITE
Feathers: Luc Hosten, Port Elizabeth (May 2003).

"I live near the sea and love watching Cape gannets. When I'm lucky they feed on passing shoals of fish in a mad, wheeling rush: birds diving from incredible heights with pinpoint accuracy and never collide. It's busier than the busiest airport. Occasionally a bird will wash up, sometimes tangled in fishing line, plastic or even those bags that oranges come in. This bird was dead, with no sign of any wounds. I guess it died of old age."

Due south: Jaco Makkink, Centurion (June 2000).

At the baobab-studded Kubu Island in the heart of Botswana's Makgadikgadi Pans. Jaco located due south (with his GPS), as it forms the centre of a circle around which the stars revolve. "I took the first (one second) exposure at 19h00 to get the dark blue colour. The second exposure (three-and-a-half hours) was taken at 19h30, when it was totally dark. I locked the shutter open with a remote release cable, and returned to the camera at 23h00, when I used a strong spotlight to illuminate the two baobab trees in the foreground."

Electricity in Pretoria: Sarel Eloff, Cape Town (July 1999).

Sarel had often tried to capture lightning storms in Johannesburg and Pretoria, but this shot is the one that really worked. He used a tripod and cable release with an exposure time of about five minutes.

PREVIOUS PAGES
Iced web: Grant Duff, Pretoria (January 2004).

It was 05h00 in the middle of winter on Dullstroom's Lakenvlei. The temperature was minus five degrees and icicles were forming on the fishing lines. "I didn't catch anything, except this picture of iced webs on the jetty." Grant always takes his camera with him fishing, and this time it really paid off.

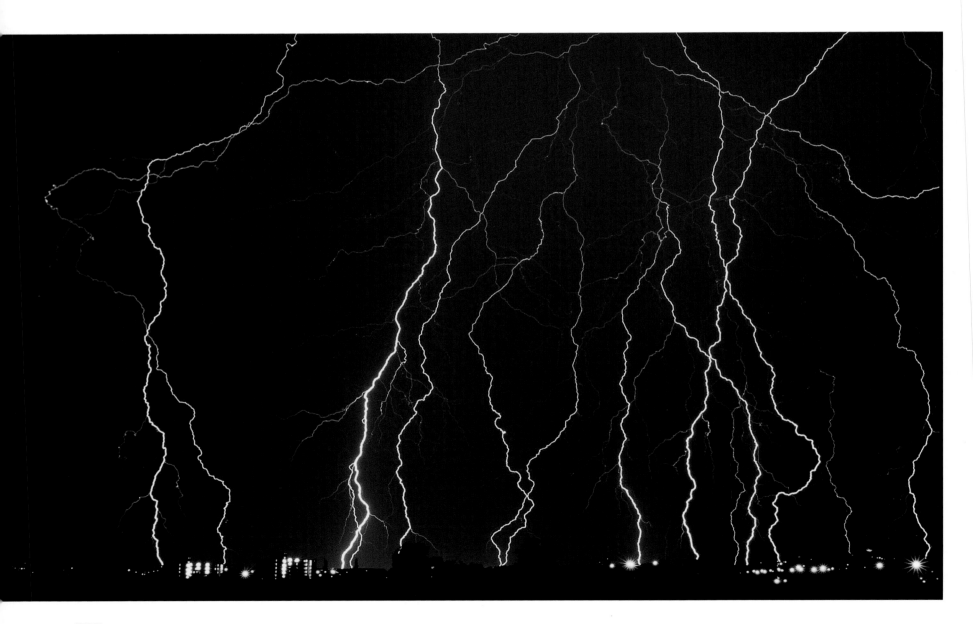

Spun spikes: Haggis Black, Adelaide (May 1999).
This spider's web on a barbed wire fence was captured early on a windless, dew-laden morning in autumn. Haggis used a 50 mm macro lens shooting into the sun. "To me, the photograph illustrates the danger of a barbed-wire fence to humans and of a spider's web to insects."

Fern on arum leaf: Friedrich von Hörsten, Somerset West (December 2002).

Friedrich is fascinated by the tints of new leaves: "The greens of oak leaves in the Cape or the range from pastel greens to plum reds of musasas in Zimbabwe." So he was relieved to find some fresh fern leaves to photograph in the otherwise dry and dull weather at Grootvadersbos. The delicate fern leaf sensuously embraced the shiny arum leaf.

Namtib star circles: Geoff Spiby, Cape Town (December 1996).

While Geoff was staying on the farm Namtib in Namibia's Tirasberg he had the chance to play with star circles time exposure. He set the aperture to f11, opened the shutter to 'bulb' and kept it open using a piece of elastic and a small stone. He then popped off two flashes (one from the left and one from the right to give a more three-dimensional effect to the tree). Two-and-a-half hours later he closed the shutter and retrieved the camera.

**Peeking prow: Peter Coqui, Port Elizabeth
(May 2002).**
It was the colours of the boat and the blue water
that attracted Peter: "I could put to good use the old
saying that less is more." The picture was taken during a South African Photographic Congress in Velddrif
on the West Coast.

Dune shadows: Ken Woods, Cape Town (October 1999).
At Sossusvlei looking across Dead Vlei Ken used a 600 mm telephoto lens. The picture is a celebration of fine lines, sharp contrasts and depth of colour, beautifully composed.

Stream lights: Ken Woods, Cape Town (May 2000).

In Namaqualand during the flower season Ken noticed these small plants growing at the edge of a stream, the late afternoon sun reflecting on the moving water. By lying on his stomach he could get the right perspective on the plants and by shooting into reflected light on the moving water he created a pleasing effect.

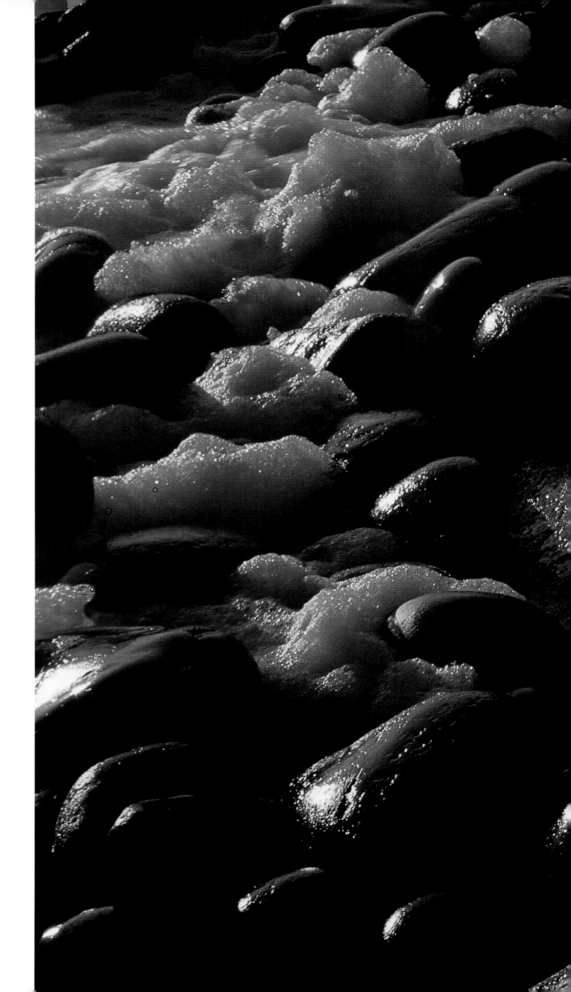

Golden rocks: Marike Bruwer, Piketberg (June 2000).

Marike spent a whole day on the West Coast shooting a boulder beach. For this picture she chose a spot where the sun's last rays bathed the rock in gold. "I underexposed by around 1,5 stops to preserve the blacks. To me, being alone in nature as the only witness to a moment like this must be one of life's greatest joys."

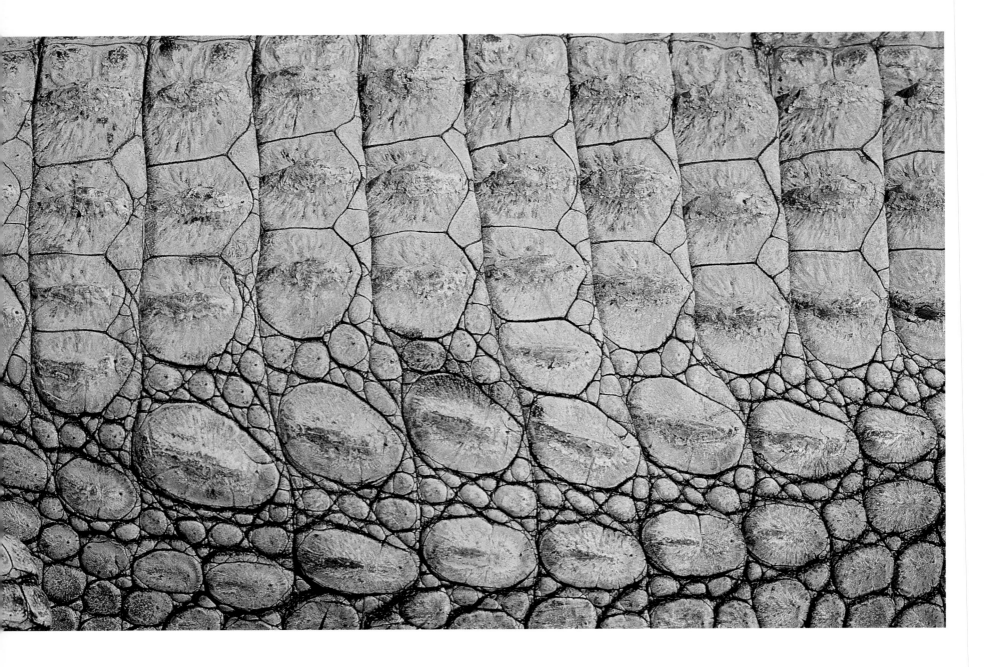

Crocodile contours: Haggis Black, Adelaide (May 1992).

We often see hexagonal building blocks in nature and it is undoubtedly one of the most structurally strong shapes. Haggis tried to illustrate this by using a long lens and keeping his composition tight. The picture was taken at a crocodile farm on the shores of Lake Kariba, so he was able to get in close. "I leant over a railing and received repeated warnings from the guide as to how high a crocodile could leap!"

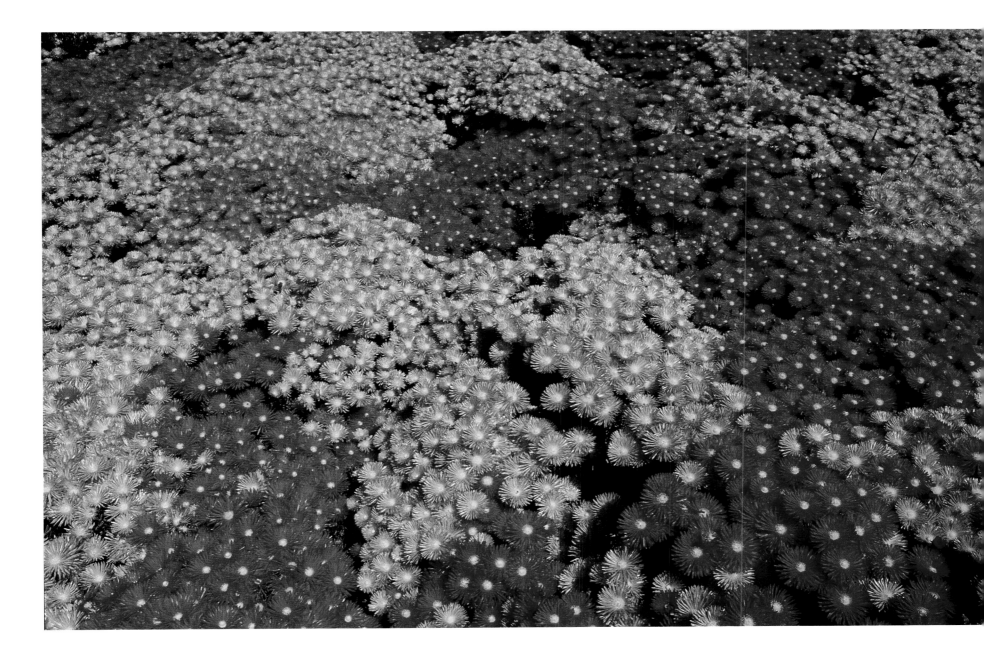

Dazzling vygies: Keith Young, Cape Town (January 2001).

Although Namaqualand is home to vygies (*Mesembryanthemums*), the Karoo Botanical Gardens near the N1 in Worcester offers a huge variety. These range from creeping groundcover and stony plants to round bushes growing nearly a metre high. This magnificent display was set against the khaki-green scrub hills of the Boland with the formidable Hex River Mountains in the background.

Boulder Bay: Willem Oets, Amanzimtoti (November 1997).

It was 40 minutes after sunset and Willem was having difficulty focusing on the rocks of the West Coast's Boulder Bay. "I was surprised to see how well the polarizing filter still eliminated the rock's reflections at this late hour. I shot at f22 with a shutter speed reading low. The misty effect of the waves moving while the shutter stayed open for about 40 seconds created a magical mood."

Richtersveld halfmens: Wayne Matthews, Kwangwanase (October 1990).

The mythical halfmens (*Pachypodium namaquanum*) of the Richtersveld is so named because it was believed to be half-human, half-plant. It grows on south-facing slopes, but turns its 'face' northwards so young leaves will catch the sun's rays.

Steps on the round:
Peter Coqui, Port Elizabeth
(November 2002).

The contrast between the handrails
and tanks at East London harbour
caught Peter's eye immediately.
Taken from a distance, he used a
tripod and f11 aperture to keep the
picture sharp.

Ship in dock: Benine du Toit, Kloof (January 1995).

This photograph was taken in the Durban harbour. "Now that you can photograph there without permits, it really is worth a visit. You just have to be aware of cranes overhead, trains shunting backwards and forwards, and the constant hive of activity. With this photograph, I loved the low light on the bow of the ship and the water running down the side. I also had fun trying slow and fast shutter speeds, either freezing the water or giving it a milky effect."

Dune climb: Wayne Matthews, Kwangwanase (December 1998).

You might say there's not much to 'Dune climb': two flat areas of dune, sky and a few tiny silhouetted figures adding interest to a strong diagonal line. But all the elements contribute to the composition. Your eye follows the hikers labouring towards the circle in the centre of the picture, which then radiates outwards to a larger circle suggested by the vignetting (usually unwanted) from the polariser.

Sickle grass: Merryl Riley, Link Hills (April 2001).

The inflorescence of this grass – seen here in KwaZulu-Natal's Krantzkloof Nature Reserve – is a one-sided spike which twists into an attractive corkscrew when dry. Merryl went in close using a 100 mm macro lens with two close-up magnifying lenses attached. She underexposed the background, while the warm backlighting created superb detail.

**Death in the desert:
Burkhard Dobiey, Windhoek
(February 2002).**

In this study of a dead gemsbok, Burkhard has explored a range of contrasts. The textures of sand, horn and bone for starters. Soft, fuzzy edges made by the sand are juxtaposed with the strong linear elements in the composition created by both the horn and the antelope's black dorsal line.

**OPPOSITE
Whirlwind in Zimbabwe:
Friedrich von Hörsten, Somerset
West (January 1996).**

August to October is dry and dusty in the Zimbabwean bush, a perfect time to photograph magnificent purple sunsets and sunrises. "But during the heat of the day everything is brown and bleak, so I usually rest up. On one such occasion this whirlwind raced through Hwange's Sinamatella Camp and showered our vehicle and cottage with leaves and dust."

Lion at Phinda: Lanz von Hörsten, Somerset West (July 2002).

This male lion was taken at Phinda Game Reserve just before sunset with a 100–400 mm lens on a Canon EOS3. The whole day had been cloudy, but Lanz saw the sun was going to break through the clouds at sunset and he asked the game guide to manoeuvre into a favourable position to wait for the shot. "The lion had just mated which is why he looks so happy. He did not have to fight to take over the pride as his father had been killed by a snare."

OPPOSITE
Male waterbuck: Philip van den Berg, Pietermaritzburg (June 1998).

Early one morning Philip was photographing in Tala Private Game Reserve in the KwaZulu-Natal Midlands. Dark clouds gathered and he thought of packing up. Then suddenly the sun broke through and the brilliant light combined with a dark background created excellent photographic opportunities, such as this shot of a beautiful male waterbuck.

Crater bulls: Steve Robinson, White River (January 2002).

Taken in the Ngorongoro Crater, Tanzania, during a time of drought, the elephant in the foreground is the largest of five bulls that are permanent residents in the crater. The unusual background colour was caused by gathering rain clouds. "The camera was a mere three metres from its weighty subject."

Ground squirrels play-fighting: Ingrid van den Berg, Pietermaritzburg (February 1999).

The Kgalagadi Transfrontier Park offers ample opportunity to watch the antics of ground squirrels. Although they spend many hours feeding, it isn't all work and no fun. From time to time they interrupt their foraging activities and engage in play sessions, chasing each other, rolling and squabbling. This captures one of those moments.

Trunk splashing: Jill Sneesby, Port Elizabeth (April 2002).
Instead of shooting the whole scene, Jill chose to go in close to accentuate the splashing action of an elephant at a Hwange water hole. This sort of behaviour can be seen almost anytime elephants go into water – the challenge was to find a new angle.

Grasshopper: Herklaas de Bruin, Pretoria (February 2003).

'Grasshopper' was taken on macro in Herklaas's garden in Pretoria North. The insect paused on a cherry tomato just long enough for him to snap the shot before it took off.

OPPOSITE
Bush snake: AJ Stevens, Swakopmund (July 1990).

While building a bush camp to film baboons in the Okavango, the crew had been warned there was a baby black mamba about. Sure enough, the shout "Mamba!" soon went up, followed by frantic evacuation. "I spotted the snake and grabbed a slender tail fast disappearing under a log. The business end then lunged at me in a coiled strike fitting of any mamba. As matters turned out, this was no mamba, but a rather irate – and harmless – spotted bush snake."

**Cape dwarf chameleon on arum:
Friedrich von Hörsten, Somerset West
(September 1999).**

Friedrich's children found this chameleon on a telephone pole in the centre of town. It was obvious the little fellow wouldn't survive for long in the traffic, so Friedrich transferred it to his garden. He couldn't resist photographing the creature on this graceful arum lily.

OPPOSITE
**Panther chameleon catching grasshopper:
Pete Oxford, Amanzimtoti (March 2001).**

Apart from lemurs, chameleons are perhaps the best-known, 'flagship' animals of Madagascar. They have radiated into dozens of species, ranging from the world's smallest to the largest. "Extremely approachable, they seem very accepting of people, perhaps because most Madagascans have an aversion to chameleons, so they are left alone."

Springbucks in mist: Johann Mader, Pretoria (September 2002).

Winter in the Kalahari can be icy cold and on some mornings mist forms in the river valleys. "While on a game drive in the Nossob area, I spotted these springboks warming up at sunrise. I liked the pattern they formed and shot the scene with backlighting to enhance their body shapes."

Ground squirrel: Johann Visser, Bloemfontein (August 1998).

One cool winter's morning in the Kalahari, Johann was lying flat on his stomach surrounded by a colony of playful ground squirrels. At first they were a bit skittish, but after a couple of hours they ventured up to him: "This squirrel was so relaxed it sat grooming itself right in front of me."

African landscapes

The African landscape begs to be rendered on film. But how to make sense of the 'chaos' of a wild vista, to find order in a scene and render this in a two-dimensional image? These are the challenges the landscape photographer faces when painting with light. Often it takes hours of patience, waiting for just the right chiaroscuro effect of sunlight dancing between shadows. Sometimes it takes days. With a cross-section of seascapes, desertscapes, forest and bush scenes, these pages show it is possible to be a Constable or Pierneef with a camera.

Khoimasis camelthorn:
Geoff Spiby, Cape Town
(February 2002).

Geoff and his family were camping on the farm Khoimasis in the Tirasberg, southern Namibia. It was late afternoon and they were rushing to reach a quiver-tree forest before sunset. As they crossed the valley floor the sun was almost on the horizon, casting long shadows. It disappeared for a moment behind this beautiful old camelthorn.

PREVIOUS PAGES
Mud tile exposure:
Hendrik Ferreira, Johannesburg
(August 1999).

During the last few seconds of sunlight on Dead Vlei at Sossusvlei you can literally see the shadows racing. Perhaps it's a metaphor for the vastness and solitude of the Namib, which is however not empty. "After several visits, I learned that the quality of the light in the Namib varies from year to year and is best during the days after the atmospheric turbulence caused by sand storms or rain."

Orange-touched windmill: Hendrik Ferreira, Johannesburg (July 2000).
"The best natural light sometimes lasts for only a few minutes: here it painted the windmill orange against dark storm clouds. We had been drenched by an unexpected thunderstorm during a hike on Weltevreden Guest Farm near Sossusvlei in the Namib. When we reached camp, the sun suddenly broke through the storm clouds."

**Crossroads: Wicus Leeuwner, Caledon
(July 1997).**

As a dairy farmer and crane conservationist living and working in the Overberg, Wicus is fortunate to have a wonderfully graphic landscape on his doorstep. "Photography has become a passion of mine and through the influence of internationally renowned photographer Freeman Patterson, I have come to view the landscape in a very different way."

Between two seas: Marike Bruwer, Piketberg (March 1996).
Taken at Dolphin's Point in Wilderness on the Garden Route, Marike had a 75–300 mm lens on her camera and was looking at the beach when she noticed a strange water pattern enveloping the fishermen every now and then. "I was a bit late with my first attempt. But after about 10 minutes of patient waiting, it happened again."

Kalahari sunstrike: Joe Lategan, King William's Town (June 1993).

Joe's image of a camelthorn in the Kgalagadi National Park has a 30-second-exposure with two flash bursts. "I concentrate on Kalahari landscapes because I have found that the Kalahari's animals make lots of photographers famous, but few photographers pay tribute to the land that makes the animals famous. I would like to spend the rest of my life there."

**The Lost City: Benine du Toit, Kloof
(February 1995).**

The fascinating architecture of the Lost City presents endless photographic opportunities. This shows the sky as it was, with no filters used. Because of the dark bottom half of the photograph, Benine was able to set a long shutter speed, giving her time to zoom out as the photograph was being recorded. For this she used a 100–300 mm Canon zoom lens.

Sossusvlei at dawn: Haggis Black, Adelaide (March 1994).

The sand dunes at Sossusvlei are one of the most photographed places in Southern Africa. "We left Sesriem camp site at 03h15 on a spring morning. Once in an elevated position, I used a 300 mm lens to concentrate on the dunes. I needed a reference point and was fortunate to have with me two teenagers who were 'rugby' fit. They ran almost a kilometre across the valley to a barcan dune."

PREVIOUS PAGES
Dust storm at Sossusvlei: Haggis Black, Adelaide (May 1998).

Soon after sunrise at Sossusvlei, Haggis noticed his lower legs were being sandblasted by the strengthening wind. Within 20 minutes he was enveloped in a small sandstorm that completely obliterated the dunes. However, in the morning light the camelthorn trees became highlighted, appearing ghostly in the barren landscape.

Skilpad Reserve: Willem Oets, Amanzimtoti (December 1994).

The Namaqualand spring of 1993 was one of particular abundance and Willem had just rediscovered the world through his new, 24 mm wide-angle lens. "I inverted the tripod's centre shaft and assumed a camera angle just centimetres above the ground. The sky polarised beautifully and the interesting cloud pattern allowed me to assign equal importance to daisies and sky."

Just the two of us:
Wicus Leeuwner, Caledon
(December 1999).

In celebration of the Overberg, Wicus has taken many photographs that capture the ever-changing patterns and colours of rolling wheat fields. These range seasonally from winter's green, new-grown shoots to the golden hues after summer harvest. "The interaction between farmers and nature leads to images that reflect the emotions and feelings of the inhabitants of this area."

Karoo trio: Hendrik Ferreira, Melville (December 2002).

In all visual arts the circle is a symbol of perfection. Here man-made circles meet nature's circle: a metaphor for sustainable use and coexistence. Sometimes the best photographs are taken after the sun has disappeared below the horizon, when pink and blue bands appear in the sky. At full moon, or a few days before, the moon is closer to the horizon at dusk, a very favourable situation for photography.

Island hopping: Lizette Elsip, Amanzimtoti (August 1997).

Lizette was out photographing early one morning on Magaruque Island when a fleet of dhows arrived from Vilankulo on the Mozambican mainland, bringing supplies and passengers. "It was quite difficult to get a pleasing composition as the beach was a hive of activity and the dhows were jostling for position. I made this image after most dhows had left and these three were the only ones remaining on the beach."

Magoebaskloof: Albie Venter, Colesberg (September 1999).

A dense mist and damp earthy smell mixed with the scent of pine needles on the Magoebaskloof Hiking Trail. Great photo opportunities presented themselves at every corner. For this shot Albie allowed a two-second exposure, resulting in a blurred walking figure and giving the picture its mystical atmosphere.

Hilton waterfall: Alan Riley, Link Hills (January 2003).

The light at Hilton Boys College estate was terrible, the waterfall was in shade and Alan was feeling totally uninspired, so he sat down and took some time just to enjoy the scene. "After half an hour of gazing at the falls, I started to appreciate what was in front of me: the glistening rocks, dancing reflections and droplets of water cascading gently down to a never-ending rhythm. A polariser removed unwanted reflections and the waterfall came alive before my eyes."

Kranskop, between Knysna and Plettenberg Bay: Keith Young, Cape Town (November 1991).

The beautiful and rugged coastal terrain of Kranskop lies in the Harkerville State Forest. A gravel road leads through indigenous forest to the first picnic site, alongside a small waterfall cascading down a sheer rock face to the Kranskop River Gorge. But it's at the second picnic site, 800 metres further on, that you get to view the splendour of Kranskop with its turquoise sea and rocky coastline, natural forest, fynbos, and hidden swimming spots. It's one of the Garden Route's gems.

OPPOSITE
Moonlight beach: Marike Bruwer, Piketberg (March 1998).

Marike and her friend Izak van Niekerk were scouting for a venue on the West Coast to hold the Photographic Society of Southern Africa's annual congress and ended up eating dinner at the Paternoster Hotel. "When we were about to go home, we saw a wonderful full moon over the bay and decided to go and 'paint' the boats with torches." Marike used a cable release that could lock.

Dwarfed against dune: Milton Evangelou, Vereeniging (June 1998).

It was late afternoon at Port Elizabeth's Maitland River Mounth and the wind had reached gale force, whipping plumes off the dune crests. Milton locked the camera securely on a tripod and used a 300 mm lens with a polarising filter.

Gemsbok, Namibia: Neil Austen, Pretoria (February 1996).

This was a chance shot taken on a private game ranch in Namibia with a herd of gemsbok disappearing over a rise in the foreground. The clarity of light and a wonderful spectrum of colours were the reward.

Mist in the Tatas Valley: Johann van Schalkwyk, Otjiwarongo (October 1991).

Tatas Mountain in the Richtersveld is a massive granite outcrop on the northern side of the road leading over Springbok Vlakte to Grasdrift – not far from the Orange River. Fog comes upriver from the Atlantic and surrounds the high peaks of Tatas Mountain whose slopes are covered with quiver trees (*Aloe pilansi*). "Of all the times we've camped, this is one of the most memorable."

Soaring wings

Bird photographers are masters of patience and pins and needles – they must be adept at climbing trees and have bodies capable of remaining curled up in uncomfortable positions for hours on end. Africa offers abundant birdlife for twitchers wielding cameras, but it is a flighty subject. No sooner does a bird appear in the camera viewfinder than it turns away or disappears. Only the fine-tuned shutter finger can freeze a bird in motion and, invariably, a very big lens. This is one area where size does count. Images range from photographing the birds in repose – trying to capture the texture of feathers and plumage, the colourful heads – to blurred images using rear-curtain flash to convey a sense of effortless flight. Some of the photographers share their secrets, but for the most part many of these pages leave you with a question which is a hallmark of fine avian photography … how did they shoot *that*?

**Spotted-back weaver:
Willem Kok, Johannesburg (December 2002).**

Males of this species perform an elaborate wing-flapping display (in order to attract females) while hanging upside down beneath the nest they're constructing. The entirely yellow crown, red eye and spotted back are the diagnostic features of the breeding male in the southern race of this species. This male's antics proved to be highly successful: within minutes of the photo being taken the nest was occupied by a female.

**PREVIOUS PAGES
Seagull flight: Ken Woods, Cape Town (July 1998).**

Ken used the combination of a slow shutter speed and flash, and hoped for the best. Often these shots don't work, but this time luck was on his side. He was sitting on a beach having evening sundowners, with gulls flying around looking for food. "This gull held its position for a moment and I got the shot. It reminded me yet again that it's always good to keep a camera at hand as you never know when the magic moment might present itself."

Pink-backed pelican: Herman van den Berg, Pietermaritzburg (May 1998).

The floods caused by Hurricane Demoina in 1984 changed the face of the Umgeni River completely. The wooded banks and islands were washed away, leaving a shallow estuary and thus forging a new habitat for birds in a densely populated urban area. The estuary is favoured not only by birds, but also canoeists. The birds became used to human activity and this created a good opportunity for photographing from a floating hide. Even the normally skittish pink-backed pelican could be approached.

Hoopoe feeding its young: Hein von Hörsten, Somerset West (April 1994).

Hein first spotted the hoopoe with food in its beak close to Helderberg near Stellenbosch and hid in a horse's stable with a pair of binoculars to locate the nest. He then erected a hide close by so the birds would get use to it. Setting up three flashes, Hein shot 'manually' without any trigger systems. "I had to pre-focus and keep my finger on the shutter button while watching the birds through a small peephole in the hide's canvas. The feeding process occurred so fast that most shots were 'misses'. After photographing for two solid days, I eventually got the pics I'd dreamed of."

Crowned crane: Ralph Paterson, Estcourt (February 2003).

Crowned cranes are usually shy and retiring wetland inhabitants. However, this individual was very curious and insisted on investigating the camera at close quarters, allowing Ralph to capture some great close ups in the crisp afternoon light.

Glossy ibis: Johann Visser, Bloemfontein (July 2000).

It was early one hot summer's morning in the Pilanesberg National Park when Johann saw this glossy ibis bathing in a little dam beside the road. Then a herd of zebras coming to drink spooked the bird and, as luck would have it, it flew up and landed on a rock right in front of Johann, where it started sunning itself. "It was too close to get a picture of the whole bird, so I took a few portrait shots."

Mozambican terns: Geoff Spiby, Cape Town (August 2001).

These terns in Mozambique were milling around, but wouldn't come close. So Geoff put the camera on a tripod, set a slow shutter speed with rear-curtain flash and a time delay of 30 seconds. "Then I walked away and the birds came back, flying around the tripod. When the camera fired, this is the image I got."

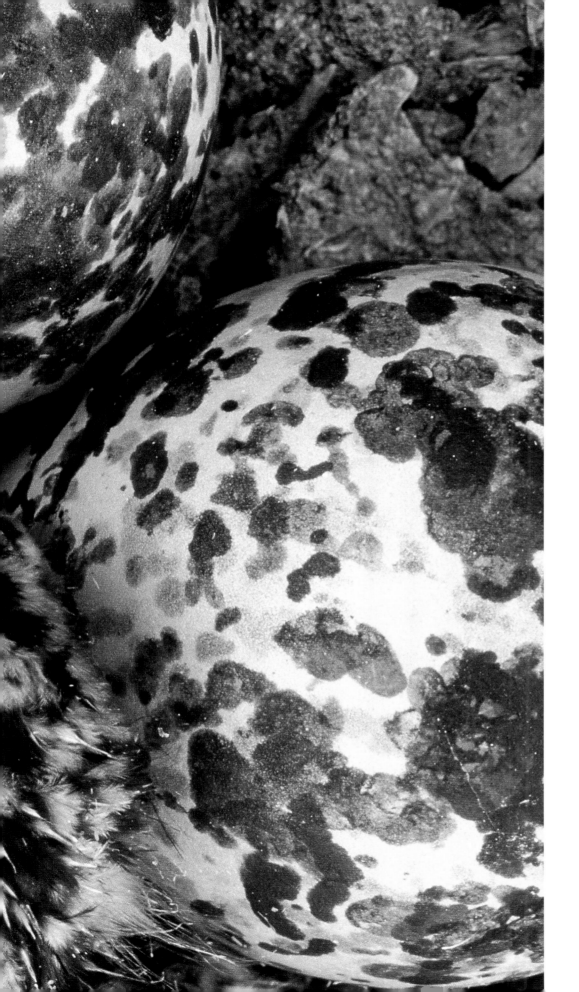

Camouflaged blacksmith plover nest: Geoff Spiby, Cape Town (July 2003).

While Geoff was on an evening game drive from his camp at Third Bridge in Moremi, a blacksmith plover appeared in front of the car. It flapped its wings and shrieked in the way they do when you get too close to a nest. "I got out the car to look. At first I thought there were four eggs, but then realised one of the 'eggs' was a chick. It was lying perfectly still so I quickly set up my tripod, shot off a few frames from directly overhead using a 105 mm macro lens, then allowed the parent back on the nest."

Photographers Index